vjbnf
631.4 PETER

VAL

Petersen, Christine
Study soils
33410016582217   06/05/20

S0-CRQ-545

Valparaiso Public Library
103 Jefferson Street
Valparaiso, IN 46383

GEOLOGY ROCKS!

STUDY SOILS

CHRISTINE PETERSEN

Checkerboard
Library

An Imprint of Abdo Publishing
abdobooks.com

ABDOBOOKS.COM

Published by Abdo Publishing, a division of ABDO, PO Box 398166, Minneapolis, Minnesota 55439. Copyright © 2020 by Abdo Consulting Group, Inc. International copyrights reserved in all countries. No part of this book may be reproduced in any form without written permission from the publisher. Checkerboard Library™ is a trademark and logo of Abdo Publishing.

Printed in the United States of America, North Mankato, Minnesota
102019
012020

THIS BOOK CONTAINS RECYCLED MATERIALS

Design: Emily O'Malley, Mighty Media, Inc.
Production: Mighty Media, Inc.
Editor: Jessica Rusick
Cover Photograph: Shutterstock Images
Interior Photographs: Adrian Wojcik/iStockphoto, p. 19; Elizabeth Dunbar/Minnesota Public Radio/AP Images, p. 11; Mighty Media, Inc., pp. 8, 9; Shutterstock Images, pp. 4, 5, 7, 13, 15, 17, 21 (all), 22, 23, 25 (all), 26, 27, 29

Library of Congress Control Number: 2019943203

Publisher's Cataloging-in-Publication Data
Names: Petersen, Christine, author.
Title: Study soils / by Christine Petersen
Description: Minneapolis, Minnesota : Abdo Publishing, 2020 | Series: Geology rocks! | Includes online resources and index.
Identifiers: ISBN 9781532191756 (lib. bdg.) | ISBN 9781532178481 (ebook)
Subjects: LCSH: Geology--Juvenile literature. | Soil mineralogy--Juvenile literature. | Soil science--Juvenile literature. | Soil biodiversity--Juvenile literature. | Pedology (Soil science)--Juvenile literature.
Classification: DDC 631.4--dc23

CONTENTS

DARWIN'S FIELD 4

A CLOSER LOOK 6

GETTING YOUR HANDS DIRTY 10

CRACKING UP 12

GETTING CARRIED AWAY 14

THE CIRCLE OF LIFE 16

LAND AND CLIMATE 18

A WORLD OF SOILS 20

LOOKING BELOW THE SURFACE 22

THE BEST SOIL AROUND 24

LIVING WITH SOIL 26

STANDING STRONG 28

GLOSSARY 30

SAYING IT 31

ONLINE RESOURCES 31

INDEX 32

DARWIN'S FIELD

In 1871, a trench was dug across a field near Charles Darwin's house. This may seem like ordinary farm work. But, Darwin was one of the world's most famous scientists. The trench was part of an experiment!

This story begins in the 1830s, when Darwin became interested in soil. Over the years, Darwin looked at numerous fields in England. He paid special attention to those left unplanted for many years.

At first, their surfaces were scattered with a layer of rocks and other natural debris. Over time, the debris appeared to sink into the ground. They still formed a layer. But, this layer was now farther under the surface! And, it was covered with a dark substance.

Charles Darwin

Scientists estimate that the earthworms in one acre (0.4 ha) of soil can produce 700 pounds (318 kg) of castings each day.

Darwin looked closely at this substance. He recognized that it was made from earthworm **castings**. There were a lot of worms in English soil. But could earthworms make enough castings to cover whole farm fields?

In 1842, pieces of chalk had been spread across that field near Darwin's house. By 1871, all of the chalk had disappeared. After the trench was dug, Darwin saw where the chalk had gone. It was buried several inches down. New castings were piled on top of it. Darwin's idea was correct. Earthworms help form soil's upper layer!

A CLOSER LOOK

Earthworms help soil form. But what is soil made of? Unlike Darwin, you don't have to wait 30 years to peek into the soil! Ask an adult to help you find a safe place to dig, such as a garden or yard. Fill a jar with soil from this place. Be careful not to harm anything living in the soil.

Find a table to work. Cover it with newspaper or plastic. Then, spread a handful of soil over the surface. Look closely. Roll the soil between your fingers. Hold it up to your nose. What do you see, feel, and smell?

Scientists know that soil has four main ingredients. These are **organic** matter, **minerals**, water, and air. Much of the organic matter is called humus. Humus comes from decayed plants and animals.

It may be hard to spot water or air in your soil sample. Yet in the ground, air fills spaces between bits of humus and minerals. Water flows into the spaces after a rain shower. Tiny water droplets cling to soil so plants have something to drink.

Minerals and organic matter affect how a soil looks and feels. ▶

TRY THIS AT HOME: EXPLORE YOUR SOIL!

WHAT YOU'LL NEED

+ soil
+ jar with lid
+ water
+ paper towels or newspaper
+ spoon

WHAT YOU'LL DO

1. Ask an adult for permission to dig up some soil.

2. Fill the jar halfway with the soil. Add water until the jar is nearly full.

3. Put on the jar's lid. Make sure it's on tight!

4. Shake the jar for about 30 seconds. The soil and water should be well mixed.

5. Let the jar sit overnight. The soil will settle and start forming layers.

6. Check out those layers! Larger particles settle at the bottom and finer particles rise to the top. What layer in your sample is thickest?

7. Do you want to explore these layers further? Spread out paper towels or newspaper on your work surface. Open up the jar and spoon out the particles floating at the top. Carefully pour out the water. Then, spoon out each layer of soil and look at it closely. Rub the soil from each layer between your fingers. Is it smooth like clay or rough like sand?

9

GETTING YOUR HANDS DIRTY

After that soil experiment, your hands will probably be dirty. Soil may be stuck under your fingernails and in the lines on your palms. But, you don't have to touch soil to get dirty.

Dirt is everywhere! Walk around on your bare feet. Play outside on a summer day. Then take a look at your skin. There it is, dirt!

Dirt and soil can both create a mess. Yet they are quite different. Put simply, dirt makes people and objects unclean. Dirt is something you want to wash off.

Soil is something we do not want to get rid of. In fact, we depend on it! Soil is a thin layer that covers much of Earth's land. It can even be found under shallow water. Soil builds up where **minerals** and humus collect. Where soil forms, plants can grow. Plants depend on soil for food and water. We depend on plants for the oxygen we breathe and the food we eat. In these ways, soil is an important **natural resource**.

A scientist who studies soil is called a pedologist. ▶

CRACKING UP

It may seem like Darwin's earthworms helped form a large amount of soil in a short time. Yet soils can take hundreds or thousands of years to develop. Five main factors lead to soil formation. These are parent material, plants and animals, land surface features, climate, and time.

Soils start forming from parent material. This is made from broken-down rocks. Rocks are made of **minerals**. So, they provide one of the four main ingredients of soil.

Some parent material is the result of **weathering**. Water, wind, and plants all cause weathering. During a storm, water may seep into cracks in rocks. If temperatures turn cold, the water freezes. Ice takes up more space than water. So, it pushes against the rock. Small cracks grow until the rock splits. Water can also **dissolve** certain minerals in rock.

Wind carries small particles. These act like sandpaper. They crash into rock and slowly wear away its surface. Even plants can cause rocks to weather. Plants often grow into cracks in rocks. Their strong roots push as they grow. This widens cracks and loosens rock.

Rock weathered by plants can become parent material. ▶

GETTING CARRIED AWAY

Weathered rocks don't always stay in place. They may be carried away to new locations by wind, water, and ice. This process is called erosion. It can add to parent material.

The power of moving water is awesome. Rainwater can rush over the land with great force. As it flows toward streams and rivers, it picks up bits of weathered rock.

The water carries sand, **silt**, and clay downstream. These lightweight sediments may be carried all the way to lakes and oceans. But if the water becomes calm, sediments settle to the bottom.

Rivers and lakes sometimes flood nearby land. In time, the water returns to its normal depth. But floods can leave behind silt, sand, clay, and **organic** matter on land. These materials become part of the soil.

Blowing wind can also pick up tiny particles loosened by weathering. Then the particles are dropped off in new places. Wherever they land, they can become parent material.

Ancient glaciers contributed to today's rich soils in places such as Norway.

Glaciers help soil formation too. As they scrape along the ground, glaciers **weather** the rock beneath them. They also carry the rock along. This eroded rock may become parent material.

THE CIRCLE OF LIFE

Have you ever left a banana sitting too long on the counter? Has the milk in your refrigerator ever gone sour? Bananas turn black and soft. Milk becomes smelly and lumpy. **Organic** materials like these don't change on their own. Bacteria and fungi cause them to decay. These tiny organisms are everywhere, including in soil.

Decay is an important natural process. Living things store **nutrients**. After plants or animals die, soil organisms break down their bodies. Some creatures chew up dead plants. Beetles and other insects eat decaying animal remains. This releases the nutrients into the soil.

You may recall that decayed plant and animal matter is called humus. This organic material is the second main ingredient of soil. It helps the soil support new plants and animals.

Look at the nutrition facts on your breakfast cereal box. You'll probably see many of the nutrients that are also found in soil. They include calcium, potassium, and iron. How did those nutrients get into your cereal? They came from soil!

Fungi will help break down these leaves so they become part of the soil.

The main ingredient in your cereal is probably wheat, corn, oats, or rice. These plants absorb water, **minerals**, and **nutrients** through their roots. They use water, air, and sunlight to make food and grow. By eating these foods, you feed your body nutrients.

LAND AND CLIMATE

You know about parent material and plants and animals. Land surface features also affect soil development. Picture a mountain with a valley below. Have you ever noticed that high mountains have few trees and plants? The soil is often thinner because soils tend to slip on steep surfaces. This erosion moves the top layer of soil down into valleys. There, thick, rich soils develop where the land is flat.

Climate is important for soil formation too. Climate affects **weathering** and plant and animal growth. If an area freezes and thaws often, weathering happens quickly. Rain can also speed up weathering as the water wears down rocks.

If the climate is harsh, the soil will not develop as much. Extreme heat sucks away moisture in soil. Freezing cold slows decay. Many plants do not grow well under extreme conditions. Without enough plants, wind blows away exposed soil.

Cold temperatures slow the decay of organic matter in northern tundra soils.

A WORLD OF SOILS

Time is the final factor in soil formation. Soil has a life cycle like people do. It starts out young and then **peaks** in quality. Later, it enters old age.

Parent material, plants and animals, land features, and climate all affect how soil ages. If humus does not form, the soil will be less fertile. But under the right conditions, the soil may have time to become deep and rich in **nutrients**.

All five soil-forming factors affect what type of soil develops. Soils occur around the world and change from place to place. There may be many types of soil across a state. In fact, there may be different types of soil in a single field! Scientists have identified more than 20,000 different kinds of soil in the United States alone.

Soils can look very different from one another. Soils can be brown, black, red, or yellow. The color tells scientists about the soil. Dark soil probably contains a lot of humus. Red soil may contain a lot of iron.

Soils come in many different colors. What does the soil look like where you live? ▶

LOOKING BELOW THE SURFACE

What does your room look like when it hasn't been cleaned for a while? Chances are, clothes pile up on the floor. Books and papers gather on the desk. You have to dig under one layer to find the next.

Soil also forms layers. Scientists call these layers horizons. They label each one with a letter. Together, soil horizons make up a soil profile.

In some soils, different horizons are well developed and easy to spot.

The **O horizon** is the soil's top layer. It forms at the surface. It is made almost entirely of humus and partly decayed plant and animal matter. The O stands for **organic**.

The **A horizon** is the next layer down. It is a mixture of humus and small bits of **minerals**. Gardeners sometimes call this dark layer topsoil. Plant roots often extend to this layer.

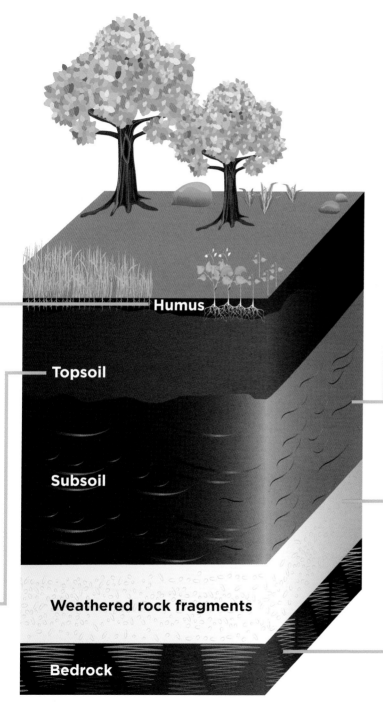

THE BEST SOIL AROUND

Soils contain different amounts of sand, **silt**, and clay. The combination of these **mineral** particles determines how a soil acts and feels. It affects how water, air, and **nutrients** move through the soil.

Sand is the largest of the three particle types. Bits of sand are rough and fairly round. They don't fit together well, so a lot of space is left between them. Water slips through these spaces.

Bits of silt and clay are much smaller than sand. Silt and clay particles are typically flat and thin, like cereal flakes. Clay can be very heavy. Its tiny particles fit together tightly and resist the flow of water. If soil has too much clay, air cannot get in. Then, plants are unable to grow.

Loam is soil with a nearly equal balance of sand, silt, and clay. The soil is rich and crumbly. Water can drain well and air can enter between the particles. Plant roots can get the water and nutrients they need to grow. Many plants grow best in loam.

Clay (*front*), loam (*center*), and sand (*back*) soils ▶

LIVING WITH SOIL

Earthworms, bacteria, and fungi are not the only organisms that live in soil. They share their surroundings with countless other important creatures. Many small animals dig and burrow into soil. This creates space for air, which can aid soil health.

Soil is an important **natural resource**. So, it needs protection from pollution and too much erosion. Pollution affects the health of soil as well as those who depend on it. Too much erosion washes away soil altogether.

Having too little **organic** matter can increase soil erosion. Luckily, many people work hard to prevent this. Ranchers do not let animals eat too much in a single field. Planting certain trees and plants also helps. Their roots help hold soil in place.

Dust storms can form when eroded soil is picked up by the wind.

COMPOST!

Compost is one way everyone can improve the soil in his or her own backyard. Gather plant parts such as grass clippings and coffee grounds. In a bin, layer them with soil and water. Leave the mixture to decay. After several months, the material will be ready to add to garden soil. Compost improves soil structure and adds **nutrients**. So, plants grow well in it.

STANDING **STRONG**

Besides being protected, soil must be studied. In 1173, construction began on a new tower in Pisa, Italy. Yet even before it was finished, problems appeared. The beautiful white marble tower began to lean.

Sometimes, buildings fall because of poor construction. But that was not the problem in Pisa. The heavy tower had been built on soil that was soft! Over the years, it leaned more and more.

Something had to be done to save the tower. In the 1990s and early 2000s, engineers propped it up with cables and weights. Then, they dug out some of the soil to make the tower lean a little less. It remains tilted. But for now, the Leaning Tower will not fall.

Modern builders try to prevent such problems by using **soil mechanics**. Before a building goes up, scientists study the soil where it will stand. Each type of soil settles differently under heavy weight. So, engineers design the best **foundation** for each type of soil. This helps buildings last and keeps people safe.

The Leaning Tower of Pisa used to lean at a 5.5-degree angle. The engineering work in the 1990s reduced the angle to 3.99 degrees.

Do you still have that jar of soil? Take it back outside and pour it in a garden. Keep an eye on it, because something may grow in it. That's what soil is all about!

29

GLOSSARY

bedrock—the solid rock beneath soil and looser rock.

casting—the waste product of an earthworm.

dissolve—to cause to pass into solution or become liquid.

foundation—the base or support upon which a building rests.

mineral—a nonliving solid with a specific chemical makeup that occurs naturally on Earth.

natural resource—a material found in nature that is useful or necessary to life. Water, forests, and minerals are examples of natural resources.

nutrient—a substance found in food and used in the body to promote growth, maintenance, and repair. The process of using nutrients is called nutrition.

organic—relating to or coming from living things.

peak—to reach the highest point.

silt—fine sand or clay that is carried by water and deposited as sediment.

soil mechanics—the study of the physical features of soil on which buildings, roads, and other structures may be built.

weathering—the breaking down of rock in its original position at or near Earth's surface. Something that is weathered has experienced weathering.

SAYING IT

fungi—FUHN-jeye

humus—HYOO-muhs

loam—LOHM

Pisa—PEE-zuh

ONLINE RESOURCES

To learn more about soils, please visit **abdobooklinks.com** or scan this QR code. These links are routinely monitored and updated to provide the most current information available.

INDEX

B
bacteria, 16, 26

C
clay, 8, 14, 24
climate, 12, 18, 20
color, 20, 21, 22, 23
composting, 27

D
Darwin, Charles, 4, 5, 6, 12

E
earthworms, 5, 6, 12, 26
England, 4, 5
erosion, 14, 15, 18, 23, 26

F
fungi, 16, 26

G
glaciers, 15, 23

H
horizons, 22, 23
humus, 6, 10, 16, 20, 22, 23

I
Italy, 28

L
land surface features, 12, 18, 20
Leaning Tower of Pisa, 28
loam, 24

M
minerals, 6, 10, 12, 17, 22, 23, 24

N
nutrients, 16, 17, 24, 27

O
organic matter, 6, 14, 16, 22, 26

P
parent material, 12, 14, 15, 18, 20
plants and animals, 6, 10, 12, 16, 17, 18, 20, 22, 24, 26
pollution, 26
protecting soil, 26, 28

R
rocks, 4, 12, 14, 15, 18, 23

S
sand, 8, 14, 24
silt, 14, 24
soil formation, 5, 6, 10, 12, 14, 15, 16, 18, 20
soil mechanics, 28
soil profile, 22

T
time, 12, 20

U
United States, 20

W
weathering, 12, 14, 15, 18, 23